Contents

Foreword
Takashi Ito

Twelve some years have passed since I became involved with Sado Island, when I "discovered" concrete facilities completely covered with grass and shrubs in the Kitazawa area. At that time, I was visiting the Oma Port constructed with pounded earth on Sado Island together with a photographer, Mr. Hoichi Nishiyama. This was the real beginning of my untying the mystery of those mining facilities. Since then, I have studied mining facilities all over Japan and have compared them with those on Sado.

Sado Island has a long mining history which dates back to the days of *Konjaku Monogatarishu* (Anthology of Tales for the Past), compiled about a thousand years ago, and such mining sites have been identified now even after such a long period of time. Although mining sites differ by age, the mines on Sado reflect a variety of mining remains from various periods in history. What really impressed me were the gold mining site which dates back to about a thousand years ago and the silver mining sites from the medieval period.

There are quite a few temples and shrines on Sado Island. This might be due to mine owners' wishes to get rich quick and become millionaires, or miners' wishes for safety while working in dark tunnels. There were almost 300 temples in existence during the Edo period (1603-1867), and 272 of them still exist now. There were also 354 shrines 250 years ago, 278 of which still exist. It's an amazing thing.

When I come to the mining sites on Sado Island, I always wonder how they pinpointed the locations of gold and silver deposits and how they dug for ores at that time. These sites remain as sanctuaries where we can think about the human desire to change world history with a fortune.

This is a collection of photographs of Sado Mine taken by Mr. Hoichi Nishiyama. What you can see in this album are mining sites and facilities which are no longer in use, since the mining operation halted in 1989. However, those remains appear lively in his photos. The atmosphere is one resulting from the accumulation of a long history, and the remains loom with overwhelming strength; the time axis and the spatial axis intertwine and speak to us. It's more than mere beauty and charm.

To me, this photo collection shows a renascence of Sado Mine by breathing new life into it.

写真集『佐渡鉱山』に寄す

文＝伊東孝

　佐渡島とは12年来の付き合いとなる。たたき工法でつくられた大間港の港湾施設を西山さんと取材したとき、樹木と草叢のなかに埋もれていた北沢地区のコンクリート施設群を発見（?）してから謎解きがはじまった。わたしが鉱山施設に取り組む原点が、ここにあった。以来、全国にある鉱山施設を訪ねては、佐渡の鉱山施設と比較した。

　佐渡の鉱山開発の歴史は、古い。『今昔物語集』にまでさかのぼることができる。今から1000年前の話だ。東京人の感覚でいえば、長い年月の間に土で覆われ、樹木が生茂り、ないしは開発されて跡形もなくなったと思っていたが、然に非ず。『今昔物語集』にはじまる平安の頃から、戦国時代、江戸時代、そして近代まで、佐渡鉱山の開発は、時代によって採鉱地域は大きく変わったが、それぞれの時代の遺跡や遺構が残っているのだ。圧巻は、1000年前の平安時代の金山跡や中世の頃の銀山跡が残されているということ。

　佐渡には社寺仏閣が多い。一獲千金を夢見た山師は、神さま仏さまに願懸けしただろうし、真っ暗な坑内で鉱石を掘っていた鉱夫たちは社寺仏閣に自らの命を託したのかも知れない。寺院数は、江戸時代約300カ寺を数えた（今日でも272カ寺が健在）。清水寺、長谷寺もある。（「せいすいじ」「ちょうこくじ」と読む。）江戸時代の神社数は宝暦期には354社あり、現代でも278社ある。

　佐渡金山のシンボルである道遊の割戸や訪れる人の少ない滝裏の間歩跡など、現地を訪れると、むかしの人（鉱夫）は、どのようにして掘ったのだろう、いやその前に、どのようにして金銀の鉱脈場所を見つけたのだろうか、と思う。金銀鉱山跡地は、世界の歴史を動かしてきた人間欲望の聖地に想いを致すことのできる場所でもある。

　本書は、西山芳一氏による佐渡鉱山の写真集である。鉱山自体は、平成元年に操業を休止し、現在あるのは、各時代の鉱山跡や使われなくなった施設や機械類などである。しかし西山さんのレンズを通すことによって、それらの遺跡や遺構にあらたな命が与えられ、彼らがざわめきはじめる。1000年の歴史の蓄積が醸し出す濃密な空間文脈や穏やかで奥深い雰囲気、圧倒的な迫力で迫ってくる遺構群など、時間軸と空間軸が絡み合いながら語りかけてくる。単に美とか魅力とか、一言で表現できうるものではない。

　本写真集には、佐渡鉱山の遺跡や遺構に、あらたな命を吹き込み、再生させる営みを見る思いがある。

Nishimikawa

Alluvial Gold Deposits 12th century-1872

The Nishimikawa area, which appears in *Konjaku Monogatarishu* (Anthology of Tales for the Past) compiled in the early 12th century, is located in south-western Sado Island and considered to be the place where the folk tale "*Kogane no Shima*" ("Island of Gold") was born. The old tale, in which alluvial gold were seen in paddy fields reflecting the rising sun, has been passed down by word of mouth in this area. The Nishimikawa area has the longest gold mining history on Sado Island. They did placer mining by digging away the surface of the land. Mountain priests from Owari in central Japan were considered to be mining in the Nishimikawa area where there is some evidence of religious activities. A mining site on Mt. Toramaruyama

still remains bald. Tools used for placer mining remain in the village of Sasagawa at the foot of the mountain. Some dams and water channels used to separate alluvial gold from gravel, stone structures considered to be used as a workshop, and numerous stones used for construction can still be seen around the mountain.

西三川砂金山　絵巻―江戸時代の砂金採取の一連の作業工程やその説明が描かれており、砂金採掘の技術を理解できる貴重な資料。
Mining picture scroll of Nishimikawa–a document in which the placer mining process during the Edo period is depicted.

西三川

西三川砂金山　平安時代末〜1872

小佐渡の西に位置する西三川地区は、『今昔物語集』（一二世紀前半）にも登場する「金の島」伝説を生んだ場所といわれる。朝日に反射してキラキラ光る砂金が、水田一面に広がっていたという。金山開発では一番古い歴史をもち、当初は地表面を削って、砂金採取をおこなった。開発には、尾張国（現愛知県）の山伏たちが関わったとされ、荒神山や阿弥陀堂など修験信仰の証も残る。山肌が削り取られた虎丸山が象徴的にそびえ、麓のひそやかな笹川集落には（土砂掘りの）鶴首・（砂と砂金を分離する）汰板などの砂金採り道具が残る。山の中を歩けば、水の勢いで石と砂金を分離するため水を貯水した受堤（ダム）跡や水路跡、コの字型の石組（作業小屋跡？）などとともに、崩れて形がわからなくなったおびただしい石がここかしこにゴロゴロしている。

P6　石組遺構―用途は不明だが、砂金採取用具の修理や置場、労働者の休憩所など、様々な用途に用いられたと考えられる。
Stone structure–considered to have been used as a shed or rest area.

P8　大山祇神社―金山の繁栄と安全とを祈願して勧請された神社で、山嶽を司る大山祇命を祭神として祀っている。
Oyamazumi Shrine–built in hope of prosperity and safety at the mine.

P9　金子勘三郎家―砂金山名主を務めた金子家の住宅で、天明年間(1781～88)の建築とされる貴重な建造物。
KANEKO Kanzaburo's house–the residence of a mining leader, built in the 1780s.

P10　虎丸山―砂金山最大の稼ぎ場であり、中腹の斜面を崩した痕跡が江戸時代の絵図のとおりに残っている。
Mt. Toramaruyama–the largest placer mining site; a vestige of mining remains on the hillside, appearing the same as it did in a map drawn in the Edo period.

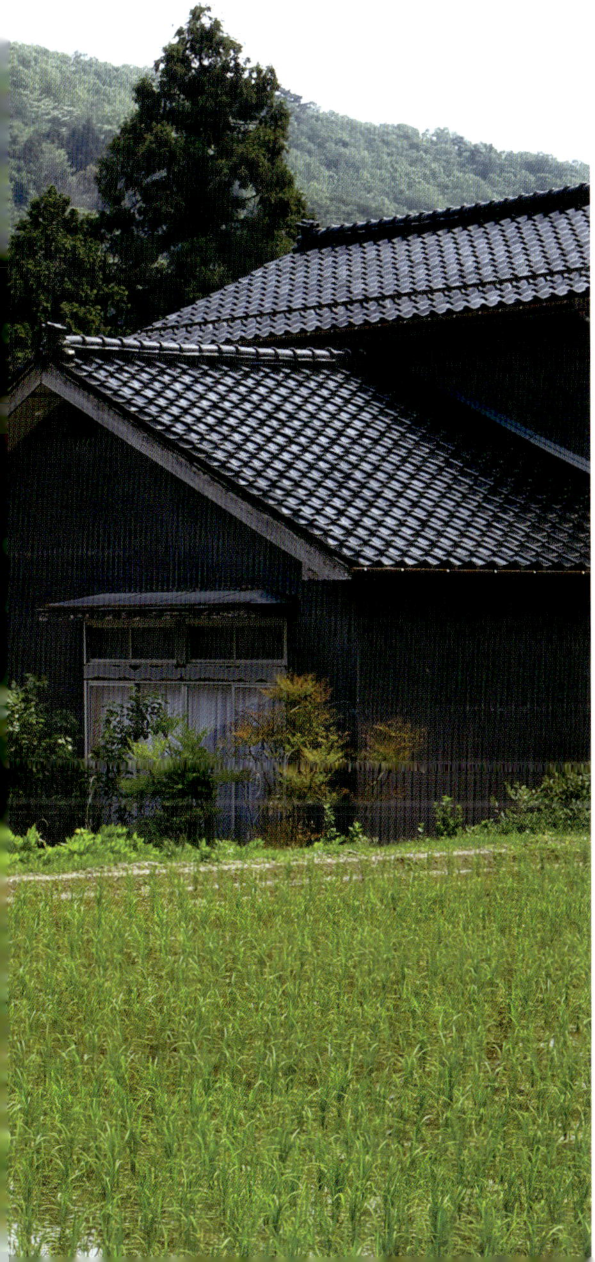

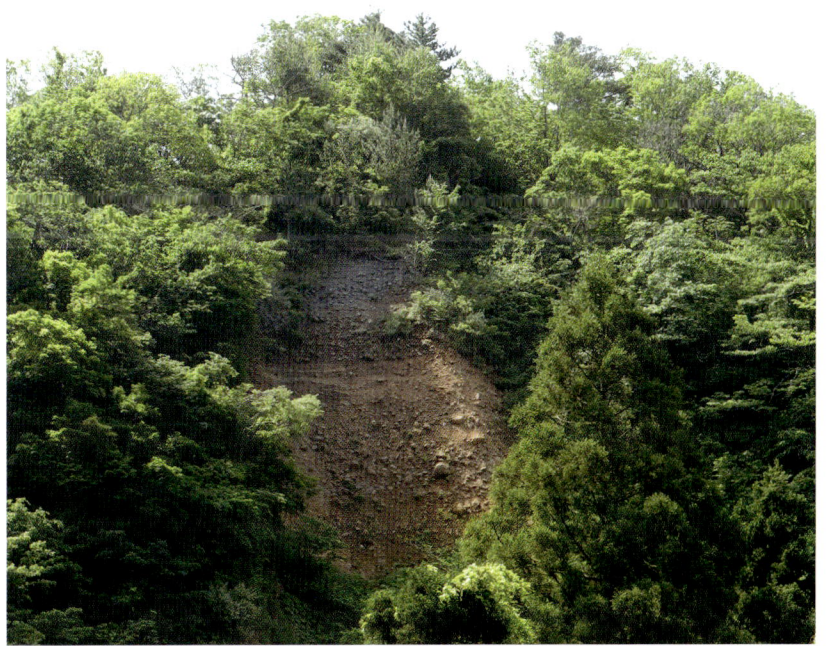

Tsurushi
Silver Mine 1542-1946

Tsurushi Silver Mine, discovered in 1542, is situated on the opposite side of Aikawa Mine. Tsurushi Silver Mine peaked in the early 17th century with thousands of residents. The smeltery was reused as a cannon foundry at the end of the Edo period. Modern mining technology was implemented after the Meiji Restoration, which led to a temporary boom in silver production. The mine operated on and off until operation was finally halted in 1946.

There are more than 600 mining sites including surface mining and tunnel mining in Tsurushi Silver Mine. The transition of mining technology is clearly visible since Tsurushi Silver Mine shows the intermediate technology between surface mining and tunnel mining. The Otaki Mabu tunnel, located at a waterfall basin, and terraces made of stones are also worth seeing.

佐渡の蝋型鋳金—初代本間琢斎が弘化４年（1847）に佐渡奉行の委託により、鶴子で大砲を鋳造したことが始まりと言われる技術。写真は６代目本間琢斎。
Wax figure casting metal–craftwork that started in 1847 with cannon founding in the Tsurushi area.

鶴子

鶴子銀山　1542〜1946

一五四二年に発見され、佐渡でも古い歴史をもつ鶴子銀山は、相川鉱山とは尾根を挟んだ反対側に位置する。最盛期は、相川の金銀山と同様、江戸初期（慶長〜寛永）の約三〇年間で、「鶴子千軒」といわれた。幕末になると、銀山の床屋（製錬所）跡は大砲鋳造所として再利用されたが、明治になって洋式鉱山技術が導入されると、一時は活況を呈した。しかしたびたび休山を繰り返し、戦後の昭和二一年に閉山した。

山の中には、たくさんの採掘跡があるが、はじめての人にはわかりにくい。しかし説明を受けて見れてくると、露頭掘りの跡や間歩などを見つけることができる。採掘法にしても露頭掘りから横掘りの坑道掘りへ至る間に、中間技術である露頭の鉱脈に沿って掘り進む鑓追い掘り跡などが残り、採掘技術の変遷が具体的にわかる。このような採掘跡が六〇〇カ所以上もあるという。中でもお勧めなのが、滝の落ち口の裏側に位置する大滝間歩跡である。周囲にある石積み擁壁でつくられたテラス遺構も見ごたえがある。

Niibo
Silver Mine 1543~1868

Niibo Silver Mine is located in the Niibo area of central Sado Island. The Niibo area, known as a farming community, is famous for Japanese crested ibis and ondeko (dancing drumming). Ondeko is considered a fusion of Noh drama and local drumming.

Niibo Silver Mine is one of the oldest mines on Sado Island. Now it is difficult to imagine that once there were more than a thousand houses in this area during its peak. Still, temples such as Kompon-ji and Seisui-ji remind us of the prosperity at that time. They are well organized and beautifully cleaned to invite visitors to a place where they can enjoy tranquility and an atmosphere rooted deeply in the continued history of the area.

新穂

新穂銀山　1543〜1868

新穂銀山が位置する旧新穂村のキャッチフレーズは、「トキと鬼太鼓」である。小佐渡の中央部にあって、村の面積を山林（六五％）と農業用地（二五％）で占めるのどかな純農村である。国際保護鳥のトキが飼育されているのもうなずける。

鬼太鼓（オンデコ）は初代佐渡奉行の大久保長安が愛好した能の影響を受けたともいわれ、独特の太鼓と村によって特徴のある振付がなされており、現在でも島全体では二二〇組もあるといわれる。

新穂は佐渡の銀山では一番古い。ここにかつては「滝沢千軒」といわれたほど民家が密集していたとは、とても信じられない。鉱山が廃止されてからの時の経過が、なせる業か。

それでも村内にある根本寺や清水寺などを訪れると、かつての栄華ぶりもなんとなく腑に落ちる。社寺境内や伽藍が、きれいに整えられ、掃除の手も行き届いている。ここは歴史が連綿と息づいている場であり、ゆっくり静謐（せいひつ）な空間と佇まいを楽しむにはもってこいの場所である。

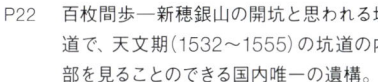

P22　百枚間歩―新穂銀山の開坑と思われる坑道で、天文期（1532〜1555）の坑道の内部を見ることのできる国内唯一の遺構。
Hyakumai Mabu tunnel–a mining tunnel which shows the configuration of the 16th century.

P23　清水寺―真言宗寺院で、境内には京都清水寺の舞台を模したといわれる救世殿（観音堂・本尊は千手観音）があり、元和8年（1622）の建立。（上）
Seisui-ji Temple–built in 1622, modeled after Kiyomizu-dera Temple in Kyoto.

P23　大日堂―大日如来を安置するが、牛の守護神でもある。険しい山道で物資を運搬する鉱山では牛が使役されたことから、牛の絵馬の献納が数多く見れる。（下）
Dainichido–the guardian god of cows is enshrined here.

Aikawa

Gold & Silver Mine 1601~1868

Aikawa Gold and Silver Mine, the largest mine on Sado Island, was discovered at the very beginning of 17th century. The Tokugawa Shogunate directly developed Aikawa Gold and Silver Mine with the most cutting-edge technology of the time. Since the Shogunate kept an isolationist policy between the beginning of the 17th century and the late 19th century, gold and silver produced in Aikawa covered its finances.

The cutting-edge technology covered all processes including ore mining, mineral dressing and smelting. Aikawa Gold and Silver Mine once had the largest output in Japan. They improved Dutch-style surveying technology for digging tunnels to make it more advanced. They also improved smelting technology and implemented "cementation process with sulfur" and "cementation process with salt". Those processes were essential for producing high-grade gold, since gold ore on Sado contained a good deal of silver.

The Minamizawa Drainage Tunnel, completed in 1696, was the result of advanced drilling and surveying technology. They started drilling from six points simultaneously to perfect the roughly one-kilometer-long tunnel. This was considered the highest level of technology at the time.

Those advanced technologies led to an abundance of gold and silver being produced.

相川

相川金銀山　1601〜1868

佐渡最大の金銀鉱山である相川鉱山は、江戸初期に発見された。幕府が直々に管理したので、当時の最先端技術を駆使して、鉱山開発を推し進めた。これによって鎖国政策をとった江戸幕府三〇〇年の財政がまかなえた。

先端技術は、「採鉱」「選鉱」「製錬」の金生産の各段階に及び、日本最大の採掘量を誇った。

坑道掘りは見えない鉱脈に向かって掘り進むので、オランダの測量技術が改良された高度な測量技術が併用された。製錬技術では、同じく石見銀山から導入された灰吹法を改良して、銀を多量に含む佐渡特有の金鉱石を製錬するため金銀吹分法と焼金法が使用されて、純度の高い金を生産した。

一六九六年竣工の排水用の南沢疎水道は、坑道掘削と測量技術の成果のひとつである。両端と中間点二本から双方に向け計六カ所から掘り進んで全長約一キロメートルのトンネルとして貫通させた技術は、当時世界最高水準といわれる。このような技術的な裏付けがあってこそ、はじめて豊富な金（銀）が産出されたのである。

P26　道遊の割戸―江戸時代に金銀鉱石を含む山を鉱脈に沿って断ち割った露頭掘りの跡で、佐渡金銀山の象徴である。
Doyu-no-warito outcrop–created as a result of mining the gold and silver vein; a symbol of Sado Gold and Silver Mine.

P28　吹上海岸石切場跡―鉱山用石磨（いしうす）の石材切り出し場で、海岸線に沿って露出する岩場に、矢穴跡や鑿跡などの痕跡が多数残っている。
Remains of Fukiage Coast Quarry–where stone materials were quarried to supply mine mortars; wedge quarry holes, traces of holes and chisel marks remain.

P29　水替無宿の墓―坑内の湧き水を手作業で汲み上げる人足として島に送られた住所不定者の墓で、火災などの坑内事故で亡くなった28名の霊が眠っている。
Tomb of homeless labors–28 labors killed in mining accidents are buried here.

P30　南沢疎水道―坑内の湧水処理のため岩盤を手掘でくりぬいた排水用の坑道。6カ所から同時に掘り進む向掘り工法により約1kmを5年という短期間で貫通させた。
Minamizawa Drainage Tunnel–a tunnel about one-kilometer-long created with only chisels and hammers to drain spring water in mining tunnels. They started drilling from 6 points simultaneously and completed it in 5 years.

P32　大安寺―徳川家康の信任の厚かった大久保長安が建立し、相川に浄土宗の信仰を広めた歴史的意義のある寺院。（上）
Daian-ji Temple–built by OKUBO Nagayasu, Sado Magistrate.

P32　大久保長安逆修塔―大久保長安が生前に自身の死後の成仏を願い建立した石塔。しかし、実際の死後は所領を没収され一族は処刑された。（下）
OKUBO Nagayasu's stone pagoda - built by OKUBO Nagayasu himself with the hope of entering Nirvana.

P33　大乗寺　円陣仏―良寛の母の生家の菩提寺で、奥の院には「ねまり遍路」と呼ばれる円陣仏があり、参れば四国遍路を果たしたと同じご利益があるといわれる。（上）
Buddha statues at Daijo-ji Temple–Buddha statues standing in a circle.

P33　法輪寺　山師味方与次右衛門の五輪塔―日蓮宗の寺院で、慶長年間に青盤間歩を稼業した山師の墓があり、元和4年（1618）の銘がみえる。（下）
MIKATA Yojiemon's pagoda at Horin-ji Temple–a mine owner's pagoda.

Sado Mine

After the Meiji era 1868-1989

After the Meiji Restoration, Aikawa Gold and Silver Mine was modernized under the name of "Sado Mine". Modern mining facilities and large machinery are well preserved in Sado Mine. Each step of the process can be seen—ore mining, crushing, dressing and smelting—from the mountain top to sea-level. Each area has its symbolic structure or landscape, which is what makes Sado Mine attractive.

Various remains such as the Odate Shaft, the Doyu-no-warito outcrop, Takato Mill and Ore Storehouse, a tool workshop, a mound, a rail, tunnels and bridges can be seen while walking along the road leading to the sea. In the Kitazawa area, large-scale concrete structures such as the Flotation and Dressing Plant and the 50-meter Thickener can be seen.

The road leads to Oma Port where there are interesting structures such as a truss bridge, cone-shaped crane pedestals and quay walls constructed by the *tataki* method, which combines rocks with well-pounded soil mixed with lime hydrate and gravel.

The Odate Area
大立地区

昭和初期の重要鉱物増産の国策により建設され、完成当初「東洋一」といわれた鉱石処理能力をもっていた北沢浮遊選鉱場

The Kitazawa Flotation and Dressing Plant was built in the 1930s under the national policy of important mineral produc enhancement. It had the highest ore processing ability in East Asia at the time.

無名異焼　北沢窯―金銀の採掘時に出る酸化鉄を多量に含んだ赤土を用い、明治時代に高温で焼成した硬質で手の込んだ焼物が完成した。
Mumyoiyaki (earthen ware)–made of red-colored iron oxide earth from the mining site.

近代

近代の佐渡鉱山　1868～1989

相川金銀山は、近代の鉱山遺構が大型機械類とともに数多く残る貴重な場所でもある。これらの遺構をめぐるには、一気に一番上の大立竪坑に向かうに限る。金鉱石の採取からはじまり、下るにつれ、破砕・選鉱、製錬されて金に至る生産工程がわかるからだ。各場所に景観的なシンボル構造物や遺構・遺跡があるのも、佐渡鉱山の大きな魅力である。さらにそれらの遺構や構造物の中を巡ることもできるのだ。

大立竪坑と坑内捲揚室、ここから佐渡鉱山のシンボル：道遊の割戸と青盤脈の鉱脈面を望む。観光坑道の鉱山入口を過ぎると、中腹には高任粗砕場と貯鉱舎、ここから坂道をのぼって機械工場．さらに山道をのぼると大立竪坑とは逆の位置に掘られた明治期の大露天掘り跡をもみることができる。鉱石運搬路の軌道盛土やトンネル・橋などを見ながら下に降りれば、北沢の選鉱場跡に至る。斜面にテラス状に構築されたコンクリート遺構とともに、直径五〇メートルのシックナー（比重差を利用した金成分の濃縮装置）が、北沢を代表する景観構造物になっている。さらに海岸に近付けば積出港になった大間港に至る。景観的に興味深いのは△トラス橋とトップがカットされた円錐状のクレーン台座だが、オタクには大間港のたたき構造物が歓迎される。

立入禁止

The Takato Area
高任地区

P52　品質の悪い鉱石を製錬した搗鉱場の跡。
The ruins of the Stamp Mill for low-grade ore.

P53　この地区唯一、明治期の構造物である石造アーチ橋。
Stone arch bridges constructed in the beginning of the
20th century.

P54 整備が進み、草木を刈られて出現した北沢浮遊選
鉱場跡。
The Kitazawa Flotation and Dressing Plant after
weeding.

P56 明治期の青化製錬所跡・昭和初期の浮遊選鉱場跡
と北沢浮遊選鉱場跡。
The Cyaniding Smeltery built in the beginning of
the 20th century and the Kitazawa Flotation and
Dressing Plant built in the 1930s.

P58 鉄筋コンクリートだけが残った北沢浮遊選鉱場跡。
The Kitazawa Flotation and Dressing Plant with its
steel rods and surrounding concrete.

P59 国内最大級の直径50mを誇るシックナーと呼ばれる
濃縮装置。(上)
The Thickener of 50-meter in diameter, one of the
largest in Japan.

P59 ライトアップされた北沢浮遊選鉱場跡。(下)
The Kitazawa Flotation and Dressing Plant lit up.

The Aikawa Area

相川地区

P66 鉱山用港湾としての基本形状が築港時のまま
残る。
Oma Port, built as a mining port with its
original construction still remaining.

P67 慶長年間から昭和まで使用された港には、
明治期のレンガ倉庫も残る。
Oma Port was used for about three hundreds
years. Brick storehouses built about a
hundred years ago can be seen.

P68 日本海の荒波を直接受けるクレーン台と
トロッコ軌道のローダー橋脚。
Crane pedestals and loader bridges, washed
by the raging waves of the Sea of Japan.

P70 夕陽も素晴らしい一級の港湾遺産。
The breathtaking scenery of the sunset at
Oma Port.

大膳神社　薪能—大久保長安によってもたらされた佐渡の能は、奉行所の奨励により急速に伝播していった。佐渡には33棟の能舞台が現存する。
Daizen Shrine, Takigi Noh (Firelight Noh)–Noh theater was introduced to Sado by OKUBO Nagayasu and spread all over Sado with strong encouragement by the Magistrate. Thirty three Noh stages still exist on the island.